The Occupation of Korea

An American Soldier's Experience

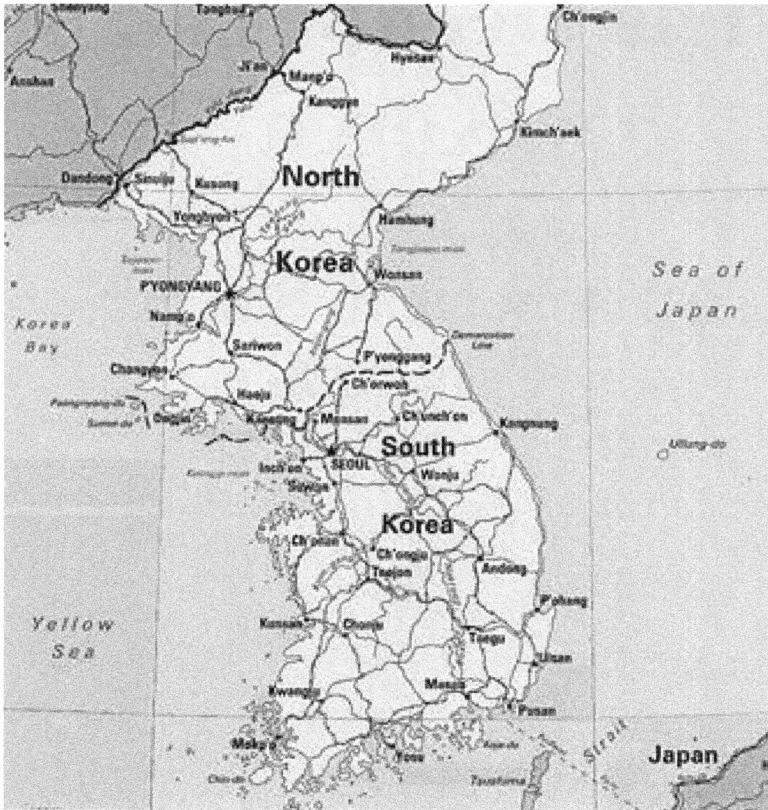

During the American Occupation of Korea (September 8, 1945 to August 15, 1948), the United States Army Military Government officially ruled over the southern half of the Korean Peninsula. The Soviet Union (now Russia) had administrative power over the northern half with a oundary line along the 38th parallel separating the two powers.

The Occupation of Korea

An American Soldier's Experience

Donald L. Stopp

W. R. PARKS
www.WRParks.com

TABLE OF CONTENTS

FOREWORD

Donald had the gift of storytelling. I say "gift" because in addition to providing facts, he made them come alive. For example, when I encouraged him to write about his experiences on board the U.S.S. Marine Swallow bound for Seoul, Korea, I not only learned that the projected fourteen-day voyage became twenty-one due to a typhoon, but also how the soap in the bathroom flew from one sink to another and sometimes landed on the floor.

Donald was a natural-born storyteller. Yet, during his lifetime, few people came to this conclusion. Everyone liked his stories, even hung on every word. But whether friends, members of writing groups, or others, they judged his writing on famous literary works. However, Don was really a storyteller, so he simply put his life memoirs down on paper.

The catalyst for his memoir writing began while participating in a writing group he attended one winter in Florida. Its goal was to create literary figures, even giants. Not until Don's death did I realize that he had been a personalized storyteller. However, many people including his mother and even professors at Duke University had long tried to make him into a literary figure, a master of the King's English. If only someone had recognized Don as a storyteller instead of trying to make him into something he was not.

Although I had yet to come to this conclusion, I doggedly encouraged his writing. What spurred me on was that everyone, including those who aspired to be published writers, enjoyed his stories so much. The distinction between them and Don was unspoken. They wrote fictional stories; he wrote about his own experiences.

Despite my myopia regarding his writing ability, a few weeks before he died in a fatal automobile accident at the age of eighty-six, he asked me if the likes of it came to pass, would I have his Korean memoir published.

What Don wrote is not the recall of an average youth just out of high school with little appreciation of the world but of a mindset that belied his years. At 18, Don already viewed Communism as an insidious threat to Democracy, or as he put it, a choice between political prison and freedom. Moreover, he was literally immersed in the American Occupation of Korea.

On behalf of Donald L. Stopp, Posthumous
Jacklin B. Stopp, Ph.D. University of Michigan
B.S. Juilliard School of Music
M.M. with Distinction, Indiana University,
Bloomington

PREFACE

In his memoir, *American Occupation of Korea*, Donald Stopp raises an important question more than once—why are we here? As a young recruit serving in the United States Army at the time of the occupation, he became increasingly interested in global history and specifically, the relationship between the U.S. and Korea and how we became involved.

Korea became a protectorate of Japan following the Korea-Japan Treaty of 1905, which gave Japan power over Korea's foreign affairs. According to the Korea-Japan Agreement of 1907, the Korean government was excluded from the administration of internal affairs. The Korean army was disbanded at this time, which helped to increase armed resistance to the Japanese that had begun after the Protectorate Treaty. The Japanese succeeded in defeating the rebels by using scorched earth tactics, but Korea never gave up striving for independence even after the country was formally made a Japanese colony in 1910.

Various groups of religious and cultural leaders signed a Declaration of Korean Independence in 1919 sparking the March First movement which began in the Korean capital of Seoul. Activists formed massive demonstrations which soon spread throughout the country in support of Korean national independence from Japan. More than two million Koreans participated in these demonstrations to publicly display their resistance and grievances against heavy taxes, forced labor, confiscation of land, and disparity in education and employment.

Korea participated in more than fifteen hundred demonstrations, but unfortunately, many were either killed or arrested. Some were executed in public as punishment. Although ineffective in accomplishing its goal of independence, the First March movement succeeded in generating an incentive for the Provisional Government Republic of Korea which was established in Shanghai one year later in April 1919.

In 1920, the Korean Independent Army was led to victory by Kim Jwa-Jin. Activist Lee Bong Chang attempted to assassinate Emperor Hirohito in Tokyo but failed to do so. In 1932, Korea dropped bombs over the Japanese military in Shanghai. By 1938, Governor-General of Korea began initiating new policies. It wasn't until 1945, however, in accordance with the terms of the Potsdam Declaration that Korea's independence from Japan seemed more likely.

The Potsdam Declaration established terms for surrender of the Empire of Japan during World War II. It was issued by U.S. President Harry S. Truman, U.K. Prime Minister Clement Attlee, and Chairman of the

Nationalist Government of China, Chiang Kai-shek. The terms were intentionally vague, thus allowing the Western Allies freedom in managing the affairs of Japan afterward and were fiercely debated as too misleading. The Japanese openly rejected it until the terms for peace were mediated by the Soviets.

President Truman warned in a widely aired speech that if Japan failed to accept the terms of the Potsdam Declaration it could "expect a rain of ruin, the likes of which have never been seen." They ignored this, however, and fought on.

The U.S. dropped its first atomic bomb on Hiroshima on August 9, 1945, and three days later another on Nagasaki. At the same time, the Soviets invaded Manchuria, and the Japanese were quickly defeated.

Prior to the end of World War II, America and Soviet diplomats had agreed to divide the Korean Peninsula at the 38th parallel with America occupying the south and Soviet forces to the north. Soon afterward, the Peoples Republic of Korea was established and grew into a centralized political party in the north, but the Americans refused to recognize the legitimacy of any political organizations in the south. It failed, therefore, in its attempt to take control over Korea and was forcibly dissolved by the U.S. Army Military Government.

Kim Gu was a leading politician who struggled for Korea's reunification and independence and was allegedly killed by the opposition. Kim Gu believed the independence of Korea as one nation was crucial to world peace. Although many Koreans campaigned for unification on the north and south, negotiations failed. The Soviets advocated a socialist form of government and the expansion of communism. The United States, however, continued striving toward the establishment of democracy and capitalism.

The US-USSR Joint Commission on the formation of the Korean Government reached impasse in 1946 after which the Cold War began, and the commission was dissolved.

In 1948, the United Nations-sponsored elections in South Korea which were held on May 10th, and Syngman Rhee was elected President of the Republic of Korea. In 1949, the Democratic People's Republic of Korea was established in North Korea, and Kim Il-sung was elected its president.

THE AUTHOR, CORPORAL DONALD T. STOPP, in SEOUL, KOREA 1947

PART I

Deployment, Arrival and Assignment

In June 1946, I graduated from high school. In August, I began basic training at Camp Polk, Louisiana as an enlisted man in the United States Army. In November, I arrived at Camp Stoneman where, in what was supposed to be sunny California, I awaited assignment to who knows where.

Eleven rainy days later, I boarded the army's Marine Swallow. After climbing many waves in a Pacific typhoon, the ship reached Inchon, Korea, but not in the expected 14 days. Instead, it took 22!

Upon disembarking, my group carried their heavy duffle bags several blocks to a waiting passenger train. In our light-weight fall jackets, we were impressed by the heavy parkas worn by the American soldiers already there. On boarding the train, we found it windowless. We were told the Japanese, who annexed Korea in 1910, took the glass before being forced to leave after World War II. Now, its occupiers were the American military.

Once the train moved, we felt like we were riding in a convertible across the North Pole. When it occasionally stopped, we quickly climbed out attempting to warm up in the dim sunlight penetrating through the clouds.

Upon reaching our assignment center, which was in Yung Dung Po, we stood outdoors in bitter weather for several hours, either at attention or at ease. After a meal, we were given two comforters and directed to a nearby factory. Placed between the machinery were cots. A pipe, about a foot above the floor, gave off a slight amount of heat, so we used it to warm our feet. Otherwise, the place was cold because overhead the windows were out.

The next day, I was among several soldiers picked up and driven to our quarters in Seoul. Twenty-four hours later, I reported to the 12th General Medical Dispensary for training as a Medical Technician. It involved working with doctors, physical therapists, and patients with minor health problems. The dispensary was also a center for tetanus, cholera, and other shots, including vaccination for small pox, required for both new arrivals to this country and for anyone departing for the states. Our patients included United States government employees and their dependents, top-level military, and people in special services.

***

Why was the United States occupying Korea?

It was a country that dated back to 2,000 B.C. However, I knew little about it.

I first heard about Korea from an American missionary who spoke at the Presbyterian Church in my hometown of Pen Argyl, Pennsylvania. His last name was

Underwood, and he and his family had been forced out of Korea after the United States declared war on Japan in 1941.

A year later, his missionary brother spoke at this church. I don't remember what either said, only that they had a great concern for the people of Korea.

Then in 1947, I would hear their mother give two lectures in a Seoul, Korea church, its denomination probably also Presbyterian.

Mrs. Horace Underwood had an amazing knowledge of Korean history and culture. She was also well informed on the topic of Korean politics because her husband, also a missionary, was an advisor to the American military governor of Korea.

A year later (1948), Mrs. Underwood came to the dispensary to see a doctor. I told her about having heard her sons when they spoke in my hometown, and now, her lectures. Pleased, she offered to show me Chosen University, which her husband's father had founded when he, too, was a missionary in Korea. Despite my best efforts to get a ride, for there was no public transportation to this university on the opposite side of Seoul from where I was stationed, I was unable to follow through. But I never forgot her invitation.

In March of 1949, by which time I was out of the service and back in the states, I learned of the death of Mrs. Underwood. While hosting a tea in her home for the faculty wives of Chosen University, two armed men suddenly forced their way in. Mrs. Underwood, a stocky woman in her fifties, physically tried to force one of them to leave. She was shot, however, and died on the way to the hospital. According to Korea's president, Syngman Rhee, the guest speaker was the Korean poet Mrs. Mo Yun Suk. As a member of a United Nations Committee, she had valuable information to contradict some Russian claims, so, the intruders were presumably communists.

***

Again, why was I, rather, my country in Korea?

During World War II, a decision was made at the Cairo Conference and reiterated at the Yalta Conference in 1945, that post-war Korea would be split along the 38th parallel between Russia and the United States.

Centuries earlier, Korea had been a vassal state to China, then, 1904 brought the Japanese occupation to keep out the Russians. A year later, Korea became a protectorate of Japan, with Japanese annexation to follow in 1910. Now, 1945 saw the division of Korea, but with 1950 to see free elections in both the North and the South.

On September 8, 1945, less than a month after Japan's surrender in World War II, the American military landed in Inchon, South Korea. They found a recently established socialistic government initiated by some

Koreans and quickly replaced it with a military command. They also found a country in crisis.

When the Japanese left, so did Korea's leadership. Koreans had not been trained to supervise agriculture, industry, sanitation, and so on. Soon, cholera broke out, unemployment became rampant, and starvation was on the horizon. Though Korea once shipped rice to Japan, the United States now had to ship this staple to Korea.

Of course, there was the Japanese literally stripping of the country. Before leaving, they printed ten times more currency than already in circulation, so their civilian and military personnel could buy anything of value to take back to Japan.

Compounding these and other difficulties was the 1945-47 migration of several million North Koreans to South Korea. This huge influx resulted in many people surviving by both buying and selling goods on the Black Market. The goods were often bought or stolen by professional thieves. In the cities, the streets became filled with roaming, homeless bands of orphaned children; while in the country, forests were stripped for wood to heat homes.

***

The People of South Korea

When I arrived in South Korea in 1946, it had about 40 million people, the North half this number. Neither part was self-sufficient. The South had light industry and agriculture, while the North had heavy industry and mining. The dams along the Yalu River provided abundant electricity, but the Americans had to buy electricity from the North to supply South Korea. Payment was in the form of American industrial equipment.

My impressions of South Koreans were that they were family oriented, even to the extent that, if you had no family, you were nobody. As for orphans, they were treated as outcasts. Also disliked were foreigners. In fact, dislike of foreigners even extended to their own merchants who traded abroad.

The individual Koreans whom I met in Seoul were either your friend or they were not. There was no middle ground. Surprisingly, all had kept their Korean culture. In fact, some old men—landlords and, as I was told, the elite of a community still followed Confucian teachings. Easily identifiable, they wore the traditional, tall black hat, and white flowing robe-like dress of the past. Otherwise, males in the city wore a blue outfit that resembled a military uniform.

Whether in the city or in the country, most women used no makeup, wore white blouses, and white knickers-like trousers. School children had uniform dress—the style and color depending on grade and school.

Among my personal contacts with individual Koreans, three stood out. Each exemplified the characteristics of being independent, industrious, and evidencing a strong sense of nationalism.

One day, I met a young man fortunate to be attending high school. He told me that he wanted to learn more English, so I agreed to tutor him. Eventually, I learned he wanted to be a physician. He arrived for a lesson one day with an arm in a cast. I asked how he broke it. He responded, "In school, we learn jujitsu." Then he proudly added, "Me number two!" I wondered if he ever became a medical doctor.

Later, another young man of high school age knocked on the door to my quarters. The building had no guard, so he simply walked in, then down a hallway to my room. When I answered the door, he asked to come inside so he could speak only with me. Our conversation was brief.

"Communism very good!" he stated. "Under communism, everybody works; not so under capitalism."

I knew he was testing me. "Do you know what prison is?" I asked.

"Yes," he said.

Then to test his English, I asked, "Do you know what jail is?"

Again, he said, "Yes."

At this point, I stated, "In prison, everybody works; communism is like that." Since there was nothing more to say, he left.

Among the Koreans who worked for the General Dispensary, there was a janitor with whom I became friendly. One day, he told me he was going to take a two-week vacation. I knew he was fortunate to have a job, so I could not imagine him taking a vacation.

"Where are you going?" I asked.

"Going to sneak into North Korea to see uncle," he explained.

Upon his return, I inquired as to what it was like to live there.

"In North Korea," he explained, "there are only government stores. You need a ration card to buy at them. You can get a card only if you are a member of the Communist Party or support it."

Then I asked, "What if you are a farmer and don't need a ration card?"

In reply, he said, "If you do not support the party, they take away your farm."

The janitor also told me that the Russians brought large numbers of Siberian families into North Korea. Because both nationalities look alike, each Siberian family was placed among twenty Korean families for a special reason. It was to see that every Korean family, from its youngest to oldest members, followed communist ways.

Sometimes when I was at work in the Dispensary, a Korean fellow around the age of eighteen would drop in to see me. He told me he had an uncle in Kansas. His visits stopped for a while, but six months later, he returned. He did not tell me where he had been; however, something about him had changed. By chance, or possibly otherwise, he reappeared on a night I had charge of the four-story building housing the dispensary.

Suddenly, the Russians unexpectedly turned off the electricity to South Korea. In a few minutes, the building's generator went on. I thought quickly and decided what I must do. I asked the Korean youth to leave and explained, "We have no communists around here." After a modest protest, he left. I never saw him again, for he apparently timed his visit merely to observe American response to the loss of electricity.

Finally, among Koreans who came to the dispensary for health clearance to go to the United States were Korea's 1936 Boston Marathon Winner and its 1948 soon-to-be-winner. Running was a national sport. Sometimes, when I was riding a jeep at the 15-mile speed limit, a long-distance Korean runner would pass by! Amazingly, even in ten-degree weather, they usually wore shorts.

Turning to non-Korean visitors to the dispensary, one day, the Russian Envoy to South Korea was seated in the waiting room at the dispensary. Because a small American boy kept running around, the envoy, appearing friendly and likable, picked him up and placed him on his lap eager to entertain the boy. The boy happily quieted down.

Several months later, the Russian envoy returned to the dispensary. By now, the Cold War had become hot, so the American boy became the personification of hostility!

Another dispensary visitor was a friend of mine from basic training. I'll call him Joe. Now, he was a Warrant Officer and lived in Korea in a private home cared for by a Korean maid, but he worked for the American army out of Tokyo. His field was intelligence. As he knew where the Russian air and naval bases were in both North Korea and in Siberia, I was impressed. Russia was a potential threat to South Korea, including its American occupiers. In fact, the Russians were already up to mischief.

Before the Japanese left Korea, they printed 20 billion won, which was ten times more than in circulation in all of Korea. When the American military arrived, they kept the won, but only for South Korean use. Newly instituted was a special currency for American servicemen. In the North, the Russians introduced their own currency, but this story does not end here.

Because the won could be used in South Korea, the Russians set up a plan. It was to give 10,000 won to any North Korean who would go into South Korea as a terrorist. Also given a gun, these men were to go to a designated town with the objective to kill its police chief, mayor, and other persons of authority. Any terrorist caught, of course, was tried by the South and executed.

During my two-year tour, dispensing of terrorists was a daily occurrence. Also, happening daily was the sabotaging of locomotives, cutting off power lines, and assassinating innocent people. In fact, killing less than one-hundred individuals in a month and sabotaging only thirty-three locomotives was considered a quiet period.

***

Demonstrations in South Korea

Several months after arriving in Korea, I had a weekday off. While in my room, I heard a loud noise, so I went outside to determine the source. My quarters sat at the top of the hill, so I could overlook the compound's wall onto the street below. There, I saw hundreds upon hundreds of Korean men noisily walking toward the boulevard that led to the Capitol Building. About two dozen spotted me, and a uniformed soldier without a gun, so, in unison, they raised their right fists and shouted at me in a foreign tongue. Since I was alone in the compound, I dashed inside, grabbed the phone and contacted the dispensary to tell them of the flow of demonstrators and that they seemed endless. Soon a 4x4 arrived to take me to the dispensary. We arrived by a back route.

Since I was off duty, I went to the roof of this four-story building and saw about 100,000 Korean men and women heading for the square bounded by City Hall, Duk Soo Palace, and the Banto Hotel. Leading this informal army were children with adults following forty abreast to form rows-and-rows of marchers. Some held banners reading, "Americans go home."

Worried the demonstrators would get out of hand, the American military soon sent in a dozen tanks, with each followed by a 4x4 truck that had a machine gun centered on its roof. Intimidated by this display of armor, courtesy of the 7th Division quartered in Seoul, the demonstrators broke rank and fled up side streets.

While there were no casualties, this was not the same for a demonstration that concurrently took place against the Russians. It happened at the North Korean capital of Pyongyang, but there the Russians surrounded 10,000 demonstrators and herded them into a concentration camp. The difference in reactions was not lost on the South Koreans.

The second demonstration I saw in Seoul, which took place about a year later, passed by my new barracks. Located on a busy main street, near a corner where two men had the job of switching tracks, so a trolley could go to the right or the left, I saw several hundred people heading for City Hall. They carried banners that read, "U.S., drive the Russians out of North Korea."

The message on the banners evidenced a transformation of Korean feeling for the United States. In fact, Korean men were soon adopting Western dress, that is, by wearing a coat, white shirt with a tie, and trousers.

***

The Russian Factor

During my tour in Korea, there was a constant worry about what the Russians might do. Along the 38th parallel, American outposts were located on one side, with Russian posts facing them within talking distance. While two Americans held this assignment regardless of their height, the Russians posted nearly a dozen of their tallest soldiers. These psychological and other kinds of warfare used by the Russians were designed to intimidate and destroy the confidence of the American troops.

According to rumor, the Americans were training 50,000 Koreans to maintain public order in the South. In the North, the Russians were supposed to be training 100,000. Moreover, they also claimed that 100,000 to 200,000 Korean troops who fought with Mao Tse Tung— the Chinese Communist leader who forced Chiang Kai-shek out of China—had joined the North Korean Army. With a possibility of facing overwhelming odds, American soldiers joked, "The Golden Gate in '48, or the salt mines of Siberia in '49." Against this backdrop, the Koreans faced the first free election in their history.

***

Politics in South Korea

Before I left Korea, about a-half-dozen men ran for president. Two stood out—Kim Soo Bok and Syngman Rhee. Based on what I learned from Korean newspapers published in English, Bok had a business following. It was of minor consideration when Rhee had been a national hero since 1897. Jailed for opposing the monarchy after being freed in 1904, he had lived out of the country. In 1919, he became president of Korea's provisional government in exile. In 1940, he lived in Washington, D. C. working for Korean independence.

After the American occupation of South Korea, Syngman Rhee returned to run for president. Soon, all the candidates received an invitation to a conference in Pyongyang, North Korea. Rhee refused to attend; he knew the nature of Russian communism. In fact, when he won the election, the Russians were so angry they turned off the electricity to South Korea!

In Seoul, despite two American ships providing some electricity to the south, it was not enough to keep the water pumps working full time. As a result, I could run water in the dispensary only fifteen minutes a day, and then it was at a designated time each evening.

Two months after Rhee won the election, I returned to the United States in July to reach the Golden Gate in '48! Discharged at Camp Stoneman, California, I never dreamed I would visit Korea fifty years later.

Author, Don Stopp, in foreground with old Banto Hotel shown in background which at one time occupied the U.S. Embassy.

PART II

The Korean Experience: Pondering the Impact

Since then I have pondered the question—can an average and introspective individual affect the policies of a country? Must a person have a focus on governments and their actions? Or is influence a random thing, since decisions are made by similar ideas, experiences, and insights by the person implementing it?

My experience in Korea certainly did affect my college studies and workings with governmental bodies. During my life journey into the Cold War, I was assigned to the Medical Clinic where I got on-the-job training to be a medical technician, and consequently, served United States government personnel, dependents, and special service military groups. I was later transferred to admissions, as well.

During this time, I realized why the United States was interested in Korea. The Yalta Agreement stated that Russia and the U.S. would occupy Korea for five years and then have free elections. Russia occupied the territory north of the 38th parallel and the U.S. south of it. The country had been occupied by the Japanese since 1904 and annexed in 1910. Prior to that, it was a hermit kingdom for a couple hundred years. It was also a vassal of China until freed by the Sino Japanese War in 1895. The United States had agreed that Korea be a protectorate of Japan in

1905. This agreement and the division of South Korea in 1945 made the U.S. somewhat responsible for Korea in a moral sense.

Occupying Korea had its problems. The Japanese had twenty billion in paper currency and distributed it to the military and its officials in Korea. They in turn bought up everything valuable and carried off whatever they could back to Japan. Yes, even the windows from the train on which we arrived.

The first landing into Korea about September 8th, 1945, was by the U.S. Navy. My brother was on board. First, some Koreans had set up a spurious socialist government to run South Korea. But the U.S. abandoned that as they were happy to occupy the country and teach the Koreans how to run a democratic government which they enjoy today.

In 1945, a series of problems arose. First an epidemic of cholera had broken out throughout Korea. The Japanese administration and supervisory personnel that controlled industry and health were gone. Wide-spread unemployment prevailed as the Japanese owned and controlled everything of great value. During World War II, Korea shipped hundreds of thousands of tons of rice to Japan. Now with the economy broken, the U.S. had to ship huge quantities of rice to Korea.

Millions of people fled from North Korea to South Korea between 1946-47 creating pressure on an economy already in great stress. These people also brought a knowledge of what the conditions of occupation were like in North Korea. The South had many homeless children roaming the streets, many stealing to survive, and a flourishing black market.

When we first arrive in Korea, geographically the North was noted for heavy industry, mining, and electricity along the Yalu River. The South was noted for agriculture and light industry and acquired its electricity from the power plants along the Yalu River in North Korea. The United States, then had to pay for the electricity for South Korea. The Russians took payment in U.S. industrial machinery.

Gradually, my interest and concern with the Cold War increased and affected my future schooling. When I was discharged from Camp Stoneman in July 1948, I traveled by train to Philadelphia, trolley to Allentown, and by bus to my hometown in Pennsylvania about twenty-three miles away.

***

Influencing Field of Study at Duke University

My high school background seemed to me inadequate as preparation for college. So, I enrolled in refresher courses at a prep school in Connecticut where I could retake subjects that would better qualify me for college. A high-school classmate had called me a jack-of-all-trades. When I went home with a Park Avenue roommate from prep school to New York City, I had to defend myself as a generalist to his father. His father, a member of the NY Stock Exchange, believed everyone should be a specialist.

One of my teachers from whom I took creative writing, physics, and solid geometry, was a gifted writer. One of his books in the 25th edition was being considered for a Hollywood movie. He had worked in administrative positions there. After reading the book, I discouraged him from having it made into a movie. It could have caused a lot of social unrest, and I said to him, "Our focus should be on Communism." His book was never made into a movie.

On March 18, 1949, a local newspaper recorded the death of Mrs. Horace Underwood. As mentioned before, two armed Koreans had broken into her home while she was having a tea for the Chosen faculty. The speaker was to be Mrs. Mo, who could make the Russian position questionable regarding the upcoming Paris Conference. Mrs. Mo disappeared when the first shot was fired, but Mrs. Underwood, a stocky woman, tried to force her assailant out the door and lost her life as a result. This murder has had a lasting impact on me.

In September 1949, I entered Duke University. I thought, perhaps, I would major in mathematics like my dad. However, two teachers, one who spoke very poor English, and the other, who had mentally already retired, cleared up any thought of mathematics. My interest in the

Cold War began to surface in my English class when I wrote a term paper on the motivations of the Russians and the Americans in Korea.

I thought I might join a fraternity like my dad, even though I did not have the money—I was on the G.I. bill. I remembered the adage—*many are called, but few are chosen.* I was to learn that my experience was not unique. The man who interviewed me for the fraternity was also being interviewed by me. He was a world-stamp collector who then specialized in American stamps. He played on the basketball team and played clarinet in the band. He conducted Boy Scout meetings and joined the Army about the same time as I did. He was at Camp Polk and on the Marine Swallow ship, and in Korea at the same time, I was. He was assigned to the medical corp. He came to Duke for the same reasons I did and majored in American history. He was a duplicate of me. But I did not join the fraternity.

In my second year at Duke, I was the co-discussion leader of the Presbyterian group. We had a speaker from the State Department who was an expert on the Berlin Blockade. However, I found two other important New York City people who thought otherwise.

During my third year at Duke, I was invited to take an American Diplomatic History course for which the University brought in a professor from Stanford University. In one session, he and I got into a debate on Korea as we had differing views on the subject. As the period ended, I let him have the last word—he said, "I was in Korea."

Sometime later, Reader's Digest came out with an article on Korea by General James Van Fleet, the military commander. He plainly stated all I had been telling the visiting professor. So, I sent the article to a highly-regarded friend in New York City who had also disagreed with me on Korea.

***

Growing Interest in the Cold War

My college career had its ups and downs, so it would be 1954 before I graduated. However, my focus on the courses I took was on the Cold War—American History, Diplomatic History, Political Science, Economics, Geopolitics, Public Administration, Physics, and Architecture. I had become a practical historian. My concern was on how the past affected the future.

It was noteworthy how I began viewing events in national and world affairs. All I was interested in was contributing to public knowledge. Any so-called effects of what I said or did were at the discretion of the reader. I believed the reason the Cold War did not become the third world war was the grace of God and the prayers of His people. For we are all conscious or unconscious instruments of His direction.

One of the classes I took was in industrial relations. One day, a teacher asked the question, "Why did the Knights of Labor fail?" I raised my hand and did not give the standard textbook answer. Instead, I stated, "They had no place in American society. For the Republican and Democratic parties were made up of a cross-section of all the people." That day, we had a substitute teacher. The regular teacher was an economic advisor to the Eisenhower administration. About a week later, President Eisenhower took up the topic of special interests.

On another occasion, my study of history showed its relationship to happenings abroad. There was a new imperialism evolving. Russia, the United States, and the United Nations were opposed to imperialism. This had some causes in World War II. Russia campaigned hard against Western imperialism. Once the former colony had broken with the mother country, sometime as soon as six months after, they would help to overthrow the new government and set up a socialist state modeled after the Soviet state, but with different government vocabulary. In my view, it was a new 20th-century imperialism with Russia as a benefactor and protector. The most advanced form would become a Russian satellite. Moscow was to become the technological capital of the world. Not knowing anyone in the State Department, I passed on information to two students who seemed to have the right contacts.

I referred to the Russian naval base on the Yellow Sea in China. I spoke about Russian Communist imperialism in China. Communist China was not recognized by the United States government until the Nixon administration, and this was in 1954. The two students were impressed.

One week later, the State Department broadcasted details about Russian Communist imperialism in China. Chinese sources picked up this news. The Chinese then demanded the Russians get out of their bases in Northern China. The Russians were so angry, they stated they would no longer ship arms to China. The Chinese following the Communist line considered all imperialism Western. They justified their action as Russian expansion. The result was that another front of United States policy of blockading Russian expansion was closed.

During my final year at Duke University, as an undergraduate, I chose a graduate course in Diplomatic History of the Far East. The professor was considered an expert on the Manchurian War of 1931 by the Japanese. They Japanese had invaded Manchuria.

One day, a Japanese professor showed up in our class. During lunch on the following day, I saw the professor looking for a seat in the crowded dining hall. I was alone, so I invited him over. I began the discussion talking about the cheap novelties Japan sent to the United States prior to World War II.

I was aware that the Japanese were sitting on their hands while dealing with the State Department. I assumed the United States government might still be having a World War II hangover, the feeling they were now a protectorate of Japan and may be treating this prideful nation equivalent to a second-class nation. I continued by saying Japan should be able to make products as good as those made in the USA or even better. This led to my main point—Japan must become a trading nation like Great Britain once was if she is to survive the Cold War.

At this point, the professor asked, "Would the United States share in the trade idea I had suggested?" Understanding the American mentality, I assured him. Then he asked, "What about Wall Street?" I told him I knew a person on Wall Street, and they would get some support there.

Then we discussed the threats to Japan from Russian and China. When I approached even the possibility of taking over or even dividing Japan, the professor assured me that would never happen. I assured him that the day will come when the Soviet Union will be defeated and many of her ill-gotten gains will be lost. I knew it was better to have Japan on our side than to have her make a separate arrangement with the Soviets or to go on her own like a loose cannon disrupting the non-Communist world.

Then we talked about the demonstrations in Japan, primarily by former Japanese soldiers converted to Communism while prisoners of Russia. There was a need to be firm and not to overreact or let it get out of hand.

I knew the Japanese needed resources for its industrial economy. This was essential for their survival as a nation. I also realized that the non-Communist world could not survive if we did not stand strongly together. We each had to stand on our own and present a united front.

At this point, the professor said to me, "You speak just like the Japanese." This meant to me that I was understanding the Japanese and getting my point across.

The professor said, "Let's go for some dessert." While we waited in line, he began saying many nasty things about the Koreans. We had not discussed them until now. He was sounding like an old Japanese warhorse. Of course, he was testing me. I shocked him with my reply when I stated, "I have a great deal of respect for the Korean people."

Upon arriving back at our table, he said, "You take care of the Russians, and we will take care of the Chinese." Then he handed me his business card. He was a professor at the University of Tokyo. No one is considered important in Japan unless he is from Tokyo. His card also revealed that he was an advisor to the Japanese Foreign Office—the equivalent of our Secretary of State. He invited me to come up to his suite and talk further. But I would never go there for I had already told him all he needed to know.

A few weeks later, that same professor gave a lecture to the students, faculty, and town people. I asked him about one of his points, whether it was strategic. I have since forgotten the point, but it was the main topic of a local Durham, NC newspaper the next day.

I read an international daily newspaper one month after I talked to the professor. The 1954 newspaper read as follows—*there is a tremendous amount of activity in the Japanese Foreign Office. They seem to have a sense of direction. They are asking to come to the United States and speak to 500 students. Secretary of State John Foster Dulles has denied their request saying, 'I make the foreign policy around here.'*

Later that year, Japan made a defense treaty with the United States. It was another front closed to Russian expansion. Also, in 1954, the Eisenhower administration created SEATO Southeastern Asia Treaty Organization primarily to block the spread of communism in that region and the Baghdad Treaty which states the U.S. would come to the aid of those countries if they were attacked by Russia. Now Russia was surrounded by treaties except for Afghanistan and India.

In May 1954, I graduated from college. This was the period of Senator Joseph McCarthy hunting down Communists by a Special Senate Committee. The McCarthy methods were creating confusion, fear, support, and indecision. He was even attacking George Marshall, former President Truman, and President Eisenhower. McCarthy had support from some leading figures.

One day in the autumn of 1954, my dad and I were raking leaves in front of our house. My dad was sweeping the sidewalk, and I was cleaning up leaves in the gutter along the curb. My dad never talked about politics or governmental affairs around me. So, I never actually knew any of his political views. Suddenly he turned around quickly and asked, "What do you think of Senator McCarthy?" He probably expected a quick emotional response, so I would express my true feelings.

Up to this time, I had never discussed the matter with anyone. But now—I thought about it and calmly replied, "Some people stand in the gutter, and they look up and see others standing on higher ground. They would reach up and pull those people down and throw them in the sewer. Once having done that, they would be standing on the higher ground, but still in the gutter." Then I continued raking the leaves.

PART III

Containing Communist Expansion

During the Kennedy administration, I was working as a caseworker. I made a critical remark about Kennedy to my colleagues. Then I suggested what we should do about a certain problem. Someone at the next table overheard me and passed on the information to my supervisor. On return to the office, my supervisor spoke up, "How dare anyone criticize President Kennedy?" One week later, Kennedy reversed his position and did just what I had suggested.

The Cuban crisis came up, and the report was that the Russians were building missile bases in Cuba that could be aimed at the United States. In discussing it with a friend, I stated, "We should immediately make a blockade of Cuba." I did not make any further suggestions. However, my friend told me he would be working in Rochester (NY) the next day and would drop into the office of the U.S. Senator in charge of Foreign Relations Committee. I felt that the State Department or Defense Department would come up with the same conclusion. I know today that John and Robert Kennedy were, at first, confused about what to do. But the administration finally blockaded Cuba and demanded the withdraw of all missiles.

The result was the Russians and the U.S. agreed to quietly take our missiles out of Turkey, then Russia would not invade Cuba. Khrushchev lost his position as chairman of the Soviet Communist Party, as a result.

In 1962, the Kennedy administration told Communist China that the United States would immediately come to the support of India. China had advanced over 200 miles inside India with its armies. Consequently, the Chinese withdrew. This statement by the U.S. administration made for a complete circle of treaties or agreements around the Soviet Union. If the United States would come to the military aid of India against China, it would do likewise against the Soviet Union. Of course, this did not cover the immediate border of Afghanistan.

Naturally, I was upset when President Kennedy had the elected government of South Vietnam overthrown. I also thought it was wrong to criticize this new government for taking wartime restrictions as we did in World War II. We should have recognized that you cannot establish an advanced democracy overnight and especially in wartime.

Obviously, American patience and foreign understanding in Asia were a low state. I did not speak out in the Kennedy or Johnson administration. Our military pacification program with a puppet government in places probably would make many Vietnamese see us not differently from how they saw the French or Japanese occupation. Our initial objective was to contain Communist expansion in Southeast Asia. It was lost in the media and military-led administrations. I found Nixon and Ford foreign policy more stable, and there was no need for me to contribute anything.

In 1980, I read a newspaper editorial opposing U.S. participation in the Moscow Olympics. I agreed with the article but felt not enough was said. So, I wrote to the State Department and listed three strong reasons why we should not participate. I included the newspaper editorial. Then I mentioned the importance of the Middle East oil to the free world. My belief was that we should make war unprofitable. Several other dangers were present. The department sent me a "thank-you" letter and referred the information to their senior members.

Shortly thereafter, my wife and I were invited to a foreign policy meeting locally. The speaker had been a missionary in Iran under the shah. I asked him about the political parties in Iran. He could not tell me, but he stated there were 33 different Communist parties in Iran. Then he led the group toward an agreement to recognize the Palestine-Liberation Organization. He had friends in the State Department and appeared to desire public support for the group. It looked like everyone was falling in line.

So, I asked the following four questions. Were the PLO leaders trained in Moscow or not? Are they not a political and economic non-entity? Do they not get a lot of their arms from the Soviet Union? Is not the only way they can pay for those arms, by following the purposes of Russian foreign policy?

Other researchers were finding a lot of written information about the PLO, and the Reagan administration was following their lead. Later, the Israel army invaded southern Lebanon and overran PLO headquarters. The record showed many had been trained in Moscow.

Once Reagan became president, he seemed to be heading in the right direction in dealing with Communism. I wrote him a letter and opened it by saying, "Thanks to you, I can sleep better at night." I felt strongly compelled to write that letter. The Cold War had been drawn out and would continue unless there was a better focus. So, it took me seventeen words to bring the series of ideas into clear attention. I had to find a new word to bring it all together. The word was *track*. The administration responded by newspaper comment, "What do you mean by *track*? Is it a goal?" Yet the way it was said one could only infer the meaning by careful reading.

Later, Oxford (dictionary) wrote to me from England about new words. As a result, a congressman quoted the first part, and President Reagan acted on the last part, which was tied to the first. The only adverse comment I received was publicly from President Nixon in a newspaper article. He said, "No one should advise the President." Today, the word *track* is in common use, such as military groups are on track for reading their goals, or my daughter is on track for a master's degree.

The last contribution came during the Nicaraguan episode. The preacher of the First Presbyterian Church of Lockport, New York traveled to Nicaragua with several other preachers. On his return, he spoke at the church before a large audience from the immediate area. He also had a long support, the head of the History Department at the University of Buffalo who delivered a speech telling how wonderful things were in Nicaragua and told of the great praise for free government medicine in that country. After the speech, a brief intermission was held followed by a question-answer period.

During the intermission, I was approached by a nurse who was politically oriented and knew me. She wondered if I would ask some questions of the speaker. I stated that what the minister saw was what the government in Nicaragua wanted them to see.

I immediately proceeded with the questioning following intermission. First mentioned was that overthrowing Somoza, the rightest dictator, only to create a leftist dictator? The two speakers agreed. Then I began stating the following—Adolph Hitler had his show pieces—the AutoBahn and full employment; Mussolini had his show piece—trains ran on time; Castro had his show piece—free medical clinics. Nicaragua also had free medicine. However, there were 500 Cuban troops reportedly there, plus 100 Bulgarian, 100 Czechoslovakian, and 100 North Korean troops present. There were also Russian military technicians working there with two Russian generals.

Then I questioned why these four military groups were now in Nicaragua. Was it to prop up the present regime? Was it to eliminate the democratic Sandinistas?

Was it to establish a base so they could expand throughout Latin America? Or was it to cut off U.S. supply lines to Latin America?

The preacher, who in the past appeared to be a Conservative, seemed to agree with me. However, the history professor waived off the idea as irrelevant. A lawyer then took up the challenge.

The results from this meeting as seen in the newspaper articles and other observations were left to the reader to decide. President Reagan spoke out against petty dictators. There was a surprise in these sources at the use of the words "show pieces". The government began to emphasize democratic solutions again in its public statements. President Reagan ordered the Naval Department to check all strategic military chokepoints around the world. One Senator spoke about expansion into Central America—this perhaps could not be related to this meeting at all, as there was infiltration into San Salvador.

PART IV

Return Visit to South Korea

In 1997, I learned about the Korean Revisit Program sponsored by the Korean Government and various civic groups. Former servicemen, accompanied by their wives or other relatives, were invited to experience a free visit to Seoul and surrounding areas. I paid the airfare for a flight on Korean Airlines for my wife and myself and spent nearly a week with twenty other Korean veterans and whoever accompanied them on this tour.

We visited Inchon to see where General Douglas MacArthur landed. While gazing at the towering statue of him, a former marine in our party recalled being the first man over the wall because he carried the radio equipment for the invasion.

We also visited a reconstructed 19th Century Korean village where we accidentally met an elderly lady who, during World War II, had been forced to be a pleasure girl for the Japanese. Having just returned to her country, we also saw her being interviewed on local television.

Then we visited Korea's extraordinary war museum which was so enormous that a half-dozen planes hung from the ceiling, while outside, even more were parked on the grounds along with large armored vehicles and weaponry. Surprisingly, of all the life-size dioramas inside the building, only a few showed American soldiers!

Standing out on our visit was the trip to Panmunjom to visit the DMZ (demilitarized zone) at the 38th parallel. Upon boarding a Humvee, we signed a paper stating if anything happened to us, insofar as, injuries, etc., we would not sue the United States government. After safely passing alongside a heavily mined area, we arrived at the Quonset building. Alternately shared by American and South Korean soldiers with those from North Korea also stationed therein, we had to be careful of our actions—so we had to keep one feisty Irish-American veteran from giving a hostile gesture to the enemy he had once fought.

Outside, as we looked across North Korea toward the city of Kaesong, we saw a model village. But it had no residents. Instead, loudspeakers bombarded South Korea with claims about the joys of living under the great leader Kim Jong. However, Koreans told us there was starvation in the North for those who did not support communism.

What a difference I observed in Seoul after a half century! Throughout the entire city, construction sites had sprung up. I even learned that 100,000 construction workers had been brought in from Southeast Asia. On the site of the old Banto Hotel, three modern hotels had been built.

In 1997, I stayed at the President Hotel while opposite it was a new building replacing the 12th Medical Dispensary where I had been assigned while serving in the United States Army. The President Hotel also faced the square bound by City Hall and Duk Soo Palace. But instead of being as I remembered it, that is, nearly empty except for an occasional ox-drawn honey wagon passing by or various kinds of military transport, there was now a busy six-lane highway!

Noticed by all the veterans was the fact that so many Korean women now appeared healthy, attractive, and dressed in stylish clothes. Gone was the drab appearance of fifty years earlier—so their world had obviously changed!

I enjoyed my second visit to South Korea—impressed by the can-do attitude of its people. In retrospect, I realized that my two-year tour there in the 1940s accounted for my great interest in the Cold War which followed when I later entered college. Even today, I view current events from the perspective of observations I made of communism while a serviceman in South Korea.

Obviously, the American Occupation did wonders for the South Koreans. To their credit, they built upon it to produce wonders for themselves!

U.S. Marines engaged in street fighting during the liberation of Seoul. M-1 rifles are carried by Marines. Dead Koreans lie in the street. M-4 "Sherman" tanks seen in the distance.

USS General George M. Randall Transport leaves Yokohama, Japan with first Korean War dead returned to the United States.

Two U.S. Navy Vought F4U-5N Corsair night fighters fly past the aircraft carrier USS Boxer (CV-21), during combat operations off Korea September 4, 1951..

Bombs dropped on supply warehouses and dock facilities at this east coast port from Fifth AirForce's B-26 bombers,

U.S. Airforce attack railrods south of Wonsan, North Korea.

In mid-1950 after the invasipon of North Korea, hundreds of thousands of South Koreans fled south.

USS Missouri fires salvos from its 16-inch guns at shore targets near Chongjin. North Korea, October 21, 1950.

Bodies of political prisoners, executed near Daejon, were discovered by South Korean soldiers.

Photo of Museum of Government General of Korea taken in Koijo (present day Seoul, Korea)

ABOUT THE AUTHOR

Donald L. Stopp

Donald Stopp grew up in Pen Argyl, Pennsylvania. Upon graduation from high school, he joined the United States Army where he was assigned to the Medical Corp stationed in Seoul, Korea (1946-48).

After being discharged from the army, he attended the Forman School in Litchfield, Connecticut in preparation for study at Duke University where he earned a BA degree in American History and subsequently attended East Stroudsburg University where he earned certification to teach in Pennsylvania public schools. Later, he became certified, as well, in New York State through courses at Canisius College and SUNY Buffalo State College.

In 1967, Don married Dr. Jacklin Bolton, an associate professor of music at the University of Buffalo. Later, Don created Timesavers Maps which he wholesaled primarily in the Northeastern states. He also started a companion retail store, Pac 'n Ship, which he later sold but remained active in the International Map Dealers Association, Dallas, Texas. He was a charter member.

As a member of the First Presbyterian Church of Lockport, New York, Don sang in the choir, as well as, its Patchwork Singers. He was also a member of the Lock City Glee Club and the Lockport Chorale. While in Florida, he was a member of the Anna Maria Island Chorus and Orchestra and the Palmetto Presbyterian Church Choir.

Don's love of music, particularly American music, resulted in his joining the Sonneck Society, presently known as the Society for American Music along with his wife Jacklin. Together, they devoted their energies to tracing the career of A. N. Johnson and through the years collected over a thousand 19th Century tune books, which are showcased in the library at the University of Maryland.

Don's interests included sports, history, and writing autobiographic articles pertaining to his childhood in Pennsylvania. Don served on the board of Niagara County Historical Society for which he prepared its Erie Canal exhibit. He was a member of the Canal Task Force and played a major role in bringing the Challenger Learning Center Space Program to Lockport, NY while serving on its board.

Donald Stopp died in a fatal automobile accident on December 15, 2014 in Palmetto, Florida.

Recent titles by William R. Parks *www.wrparks.com*
Available at *Amazon.com* or wherever books are sold:

We Remember the Day of President Kennedy's Assassination

Made in America

The Joyful Cook's Guide to Heavenly Greek Cuisine

Boolean Algebra and Switching Circuits

Computer Number Bases

Introduction to Logic

Sets, Numbers and Flowcharts

Beginning Algebra

Handbook for Piano Practice

Piano Practice for the Advancing Student

Letters to a Young Math Teacher

Program Your Calculator

A Franciscan Odyssey

Windows to Heaven

Prayers from the Heart

The Nature Watch Collection Book One

The Nature Watch Collection Book Two

The Calico Caterpillar

Animal Tails

Peppin Puffin to the Rescue

Choices

Time to Fly